INTERNATIONAL CENTRE FOR MECHANICAL SCIENCES

COURSES AND LECTURES - No. 75

MORTON E. GURTIN
CARNEGIE-MELLON UNIVERSITY - PITTSBURGH

# ON THE THERMODYNAMICS OF ELASTIC MATERIALS AND OF REACTING FLUID MIXTURES

COURSE HELD AT THE DEPARTMENT
OF MECHANICS OF SOLIDS
JUNE 1971

UDINE 1971

SPRINGER-VERLAG WIEN GMBH

This work is subject to copyright.

All rights are reserved,

whether the whole or part of the material is concerned

specifically those of translation, reprinting, re-use of illustrations,

broadcasting, reproduction by photocopying machine

or similar means, and storage in data banks.

© 1972 by Springer-Verlag Wien

Originally published by Springer-Verlag Wien-New York in 1972

ISBN 978-3-211-81178-8     ISBN 978-3-7091-2874-9 (eBook)
DOI 10.1007/978-3-7091-2874-9

# PREFACE

These lecture notes offer a modern, concise treatment of two important topics in modern continuum thermodynamics: the thermodynamics of elastic materials and the thermodynamics of chemically reacting fluid mixtures.

Udine, July 1971

## Introduction

These lecture notes are not meant to be complete in any sense. They simply offer a modern, concise treatment of two important topics in continuum thermodynamics.

### Notation.

$R$ = reals $(\alpha, \beta, \gamma, \ldots \in R)$  $R^+ = \{\alpha \in R : \alpha > 0\}$

$E$ = three-dimensional Euclidean space $(X, Y, \ldots, x, y, \ldots, \in E)$

$V$ = associated vector space $(u, v, w, \ldots \in V)$

$u \cdot v$ = inner product of $u, v$

$L$ = set of all second-order tensors (tensor = linear transformation from $V$ into $V$) $(F, L, S, T, \ldots \in L)$

$\quad\quad tr S \quad\quad$ = trace of $S$

$\quad\quad S^T \quad\quad$ = transpose of $S$

$\quad\quad \det S \quad\quad$ = determinant of $S$

$\quad\quad S \cdot T \quad\quad$ = $tr(S^T T)$ = inner product of $S, T$

$\quad\quad 1 \in L \quad\quad$ = unit tensor, $0 \in L$ = zero tensor

$L^+ = \{F \in L : \det F > 0\}$

$sym L = \{S \in L : S = S^T\}$.

Given a motion $(X, t) \rightarrow x(X, t)$ of a body $B$ and a function $(X, t) \rightarrow f(X, t)$:

$\nabla f$ = material gradient (with respect to X holding t fixed)

$\dot{f}$ = material time derivative (with respect to t holding X fixed)

$\text{grad } f$ = spatial gradient (with respect to $x = x(X,t)$ holding t fixed)

$\text{div } f$ = spatial divergence (with respect to $x = x(X,t)$ holding t fixed)

# Chapter 1

# THERMODYNAMICS OF ELASTIC MATERIALS

## 1.1. Basic Laws

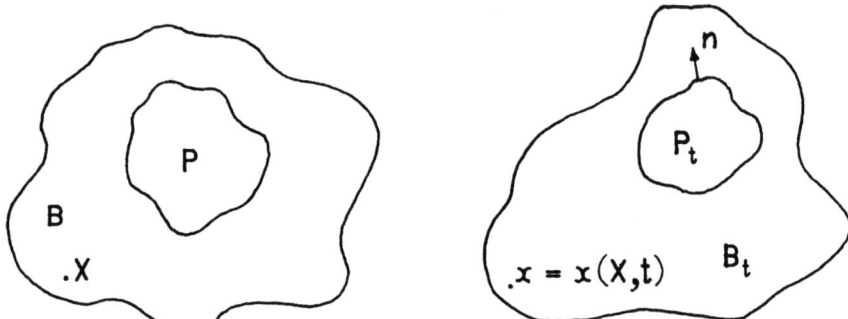

We identify the body B with the compact regular region in E that it occupies is a fixed reference configuration. A part P of B is a regular subregion of B.

<u>Definition.</u> Let T be a closed interval in R and let P be a part of B. A <u>motion</u> of P during the time interval T is a class $C^2$ mapping

$$x : P \times T \longrightarrow E$$

such that $X \longrightarrow x(X,t)$ is a smooth homeomorphism of P for every $t \in R$.

# Basic Laws

$x = X(X,t)$ is the place occupied by the material point $X \in P$ at time $t \in T$ ; $P_t = x(P,t)$ is the region occupied by $P$ at time $t$ ; $F = \nabla x$ is the deformation gradient; $v = \dot{x}$ is the velocity.

## Terminology.

$T(X,t)$    Cauchy stress tensor $(\in \text{sym } L)$
$b(X,t)$    body force $(\in V)$    (includes inertial body force)
$\mathcal{E}(X,t)$    internal energy $(\in R)$
$\eta(X,t)$    entropy $(\in R)$
$\theta(X,t)$    temperature $(\in R^+)$
$q(X,t)$    heat flux $(\in V)$
$\gamma(X,t)$    heat supply $(\in R)$
$\varrho_0(X)$    density in reference configuration $(\in R)$
$\varrho(X,t)$    current density $(\in R)$

$$\varrho |\det F| = \varrho_0 . \tag{1.1.1}$$

## Basic Laws.

### Balance of forces :

$$\int_{\partial P_t} Tn\, da + \int_{P_t} b\, dm = 0 ,$$

### Balance of Energy :

$$\frac{d}{dt}\int_{P_t} \mathcal{E}\, dm = -\int_{\partial P_t} q\cdot n\, da + \int_{P_t} \gamma\, dm + \int_{\partial P_t} v\cdot Tn\, da + \int_{P_t} v\cdot b\, dm .$$

Growth of Entropy :

$$\frac{d}{dt}\int_{P_t}\eta\, dm \geq -\int_{\partial P_t}\frac{q\cdot n}{\theta}\, da + \int_{P_t}\frac{\gamma}{\theta}\, dm.$$

Here $dm = \varrho\, dv$, and these laws are required to hold for every part P.

If the above fields are sufficiently smooth, these relations are equivalent to

(1.1.2) $\quad\quad\quad\quad\quad \text{div}\, T + \varrho b = 0,$

(1.1.3) $\quad\quad\quad\quad \varrho \dot{\varepsilon} = -\text{div}\, q + \gamma + T\cdot\text{grad}\, v,$

(1.1.4) $\quad\quad\quad\quad\quad \varrho \dot{\eta} \geq -\text{div}\left(\frac{q}{\theta}\right) + \frac{\gamma}{\theta}.$

Exercise. Derive (1.1.2) - (1.1.4).

We define the free energy $\Psi$ and the Piola-Kirchoff stress tensor $S$ by

(1.1.5) $\quad\quad\quad\quad\quad \Psi = \varepsilon - \theta\eta$

(1.1.6) $\quad\quad\quad\quad\quad S = T(F^T)^{-1}.$

(Actually, $(\det F)^{-1}S$ is the Piola-Kirchoff stress tensor). Then (1.1.3) and (1.1.4) yield the <u>reduced dissipation inequality</u>:

(1.1.7) $\quad\quad\quad\quad \varrho(\dot{\Psi} + \eta\dot{\theta}) - S\dot{F} + \frac{1}{\theta}q\cdot g \leq 0,$
where

(1.1.8) $\quad\quad\quad\quad\quad g = \text{grad}\,\theta$

is the temperature gradient.

Processes                                                                9

---

Exercise. Show that granted (1.1.2), (1.1.3), (1.1.5), (1.1.6), (1.1.8), then (1.1.4) $\Leftrightarrow$ (1.1.7).


Definition. Let P be a part and T a time interval. A <u>process</u> or P during T is an ordered array $p = [x,\theta,\Psi,S,\eta,q]$ such that

$x$ is a motion of P during T,

$\theta: P \times T \rightarrow R^+$ is class $C^2$,

$$\left.\begin{array}{l} \Psi: P \times T \rightarrow R \\ S: P \times T \rightarrow L \\ \eta: P \times T \rightarrow R \\ q: P \times T \rightarrow V \end{array}\right\} \text{ are class } C^2,$$

Remark. Given a process $p$, (1.1.2) and (1.1.3) (with the aid of (1.1.5) and (1.1.6)) can be used to compute the body force $b = b(p)$ and the heat supply $\gamma = \gamma(p)$ necessary to support $p$.


Exercise. Show that for a process $p$ consistent with (1.1.7) if $\dot{F} = g = 0$, then

$$\dot{\Psi} \leq 0 \quad \text{when} \quad \dot{\theta} = 0$$
$$\dot{\varepsilon} \leq 0 \quad \text{when} \quad \dot{\eta} = 0$$
$$\dot{\eta} \geq 0 \quad \text{when} \quad \dot{\varepsilon} = 0$$

## 1.2. Elastic Materials

An elastic material is defined by constitutive relations of the form

(1.2.1)
$$\Psi = \hat{\Psi}(F,\theta,g,X)$$
$$S = \hat{S}(F,\theta,g,X)$$
$$\eta = \hat{\eta}(F,\theta,g,X)$$
$$q = \hat{q}(F,\theta,g,X).$$

More precisely, $\Psi(X,t) = \hat{\Psi}(F(X,t),\theta(X,t),g(X,t),X)$, etc. The functions $\hat{\Psi},\hat{S},\hat{\eta},\hat{q}$ are called <u>response functions</u>; their domain is

$$D = L^+ \times R^+ \times V \times B.$$

We assume that the response functions are smooth, and that

$$\hat{S}F^T = F\hat{S}^T.$$

<u>Remark</u>. Henceforth we suppress the argument $X$ in (1.2.1)

<u>Definition</u>. A process $p$ consistent with (1.2.1) is called a <u>constitutive process</u>.

<u>Definition</u>. A pair $[x,\theta]$ with $x$ a motion of $P$ during $T$ and $\theta: P \times T \rightarrow R^+$ a class $C^2$ map is called an <u>admissible pair</u> for $P$ during $T$. Given an admissible pair $[x,\theta]$ there exists a unique constitutive process $p = p(x,\theta) = [x,\theta,\Psi,S,\eta,q]$ for $P$ during $T$. $p$ is the constitutive process <u>generated by</u> $[x,\theta]$.

**Theorem.** Necessary and sufficient that every constitutive process be consistent with the reduced dissipation inequality (1.1.7) is that the following three statements be true:

(i) $\hat{\psi}, \hat{S}$ and $\hat{\eta}$ are independent of $g$, i.e.
$$\psi = \hat{\psi}(F,\theta), \quad S = \hat{S}(F,\theta), \quad \eta = \hat{\eta}(F,\theta). \quad (1.2.2)$$

(ii) $\hat{\psi}$ determines $\hat{S}$ and $\hat{\eta}$ through the relations
$$\hat{S} = \varrho \hat{\psi}_F, \quad \hat{\eta} = -\hat{\psi}_\theta. \quad (1.2.3)$$

(iii) $\hat{q}$ obeys the heat conduction inequality
$$\hat{q}(F,\theta,g) \cdot g \leq 0 \quad \forall (F,\theta,g) \in D. \quad (1.2.4)$$

**Proof.** By the chain rule, for every admissible pair $[x,\theta]$
$$\dot{\psi} = \hat{\psi}_F(F,\theta,g) \cdot \dot{F} + \hat{\psi}_\theta(F,\theta,g) \dot{\theta} + \hat{\psi}_g(F,\theta,g) \cdot \dot{g}.$$

Thus a constitutive process $p$ is consistent with (1.1.7) if and only if

$$[\varrho \hat{\psi}_F(F,\theta,g) - \hat{S}(F,\theta,g)] \cdot \dot{F} + \varrho[\hat{\psi}_\theta(F,\theta,g) + \hat{\eta}(F,\theta,g)]\dot{\theta} +$$
$$+ \varrho \hat{\psi}_g(F,\theta,g) \cdot \dot{g} + \frac{1}{\theta} \hat{q}(F,\theta,g) \cdot g \leq 0. \quad (1.2.5)$$

Clearly (i), (ii), (iii) imply (1.2.5). To prove the converse assertion assume that (1.2.5) holds for every admissible pair $[x,\theta]$ (and hence for every constitutive process). Choose $(F_0, \theta_0, g_0, X_0) \in D$, $A \in L$, $\alpha \in R$, $a \in V$, and let

$$x(X,t) = x_0 + (F_0 + tA)[X - X_0],$$
(1.2.6)
$$\theta(X,t) = \theta_0 + t\alpha + [g_0 + ta] \cdot [x(X,t) - x_0].$$

Then there exists a $\tau > 0$ and a part $P$ containing $X_0$ such that $[x, \theta]$ is an admissible pair for $P$ during $T = [-\tau, \tau]$. Further at $X = X_0$, $t = 0$,

$$F = F_0, \quad \theta = \theta_0, \quad g = g_0, \quad \dot{F} = A, \quad \dot{\theta} = \alpha, \quad \dot{g} = a;$$

therefore, if we apply (1.2.5) to $[x, \theta]$ defined in (1.2.6) and use the fact that $F_0, \theta_0, g_0, X_0, A, \alpha, a$ are arbitrary, we arrive at (i)-(iii).

For the remainder of this section we assume that (i) – (iii) hold. The result of (1.2.3) implies the Maxwell relation

(1.2.7)
$$\hat{S}_\theta = -\varrho \hat{\eta}_F.$$

In addition, given any constitutive process $p$ we have the Gibbs' relations

(1.2.8) $\qquad \dot{\psi} = \frac{1}{\varrho} S\dot{F} - \eta\dot{\theta}, \quad \dot{\varepsilon} = \frac{1}{\varrho} S\dot{F} + \theta\dot{\eta},$

where $\varepsilon$ is given by (1.2.5). By $(1.2.8)_2$, (1.2.6), and (1.2.3); if a constitutive process $p$ is adiabatic (i.e. $q \equiv 0$, $\gamma(p) = 0$), then

$$\dot{\eta} \equiv 0.$$

Exercise. Establish (1.2.8)

In view of (1.1.5), we define the response function $\hat{\varepsilon}$ for the internal energy by

$$\hat{\varepsilon}(F,\theta) = \hat{\psi}(F,\theta) + \theta\hat{\eta}(F,\theta) . \qquad (1.2.9)$$

We call

$$c(F,\theta) = \hat{\varepsilon}_\theta(F,\theta) \qquad (1.2.10)$$

the specific heat. By $(1.2.3)_2$, (1.2.9), and (1.2.10),

$$c(F,\theta) = \theta\hat{\eta}_\theta(F,\theta) . \qquad (1.2.11)$$

We call

$$K(F,\theta) = -\hat{q}_g(F,\theta,g)\big|_{g=0} \qquad (1.2.12)$$

the conductivity tensor.

Theorem.

(i) The heat flux vanishes when the temperature gradient vanishes:

$$\hat{q}(F,\theta,g) = 0 \quad \text{when} \quad g = 0 . \qquad (1.2.13)$$

(ii) $K(F,\theta)$ is positive semi-definite.

(iii) Let $\delta = |F-1| + |\theta - \theta_0| + |g|$, and let $K_0 = K(1,\theta_0)$
Then

$$q = \hat{q}(F,\theta,g) = -K_0 g + o(\delta) \quad \text{as} \quad \delta \to 0 ;$$

i.e. Fourier's law holds to within an error of $o(\delta)$.

**Proof.** Fix $(F,\theta)$, and let $\hat{q}(g) = \hat{q}(F,\theta,g)$, $K = K(F,\theta)$.
Since $\hat{q}$ is smooth,

$$\hat{q}(g) = \hat{q}(0) - Kg + o(|g|) \quad \text{as} \quad g \to 0,$$

and thus, by (1.2.4),

$$\hat{q}(g) \cdot g = \hat{q}(0) \cdot g + g \cdot Kg + o(|g|^2) \leq 0.$$

This inequality implies (i) and (ii). To establish (iii) note first that, by (1.2.13),

(1.2.14) $\quad \hat{q}_F = \hat{q}_\theta = 0 \quad \text{when} \quad g = 0 ;$

if we expand $\hat{q}$ in a Taylor series about $(F_0,\theta_0,0)$ and use (1.2.13) and (1.2.14), we arrive at (iii).


**Exercise.** An <u>elastic fluid</u> is defined by constitutive relations of the form

(1.2.15)
$$T = -p1$$
$$p = \hat{p}(v,\theta)$$
$$\Psi = \hat{\Psi}(v,\theta)$$
$$\eta = \hat{\eta}(v,\theta)$$
$$q = -k(v,\theta)g$$

where $T = SF^T$ is the Cauchy stress, $p$ is the pressure, $k(v,\theta)$ is the (scalar) conductivity, $v$ is the specific volume :

$$v = \frac{1}{\varrho} = \frac{1}{\varrho_0}|\det F|.$$

Show that, as a consequence of the results of this section,

$$\hat{p} = -\hat{\psi}_v, \quad \hat{\eta} = -\hat{\psi}_\theta, \quad k \geq 0.$$

Hint : Use the identity $\overline{\det F} = (\det F)\,\text{tr}(\dot F F^{-1}) = (\det F)\cdot$ $\cdot \text{div}\, v$ to establish the result

$$S\cdot \dot F = -\frac{1}{v}p\dot v,$$

and then use the Gibbs' relation (1.2.8).

Exercise : An elastic fluid with $(1.2.15)_2$ in the form

$$pv = R\theta \quad (R = \text{scalar constant})$$

is called a <u>perfect gas</u>. Show that for a perfect gas $\hat{\varepsilon}$ is independent of $v$ ; i.e.

$$\varepsilon = \hat{\varepsilon}(\theta).$$

## Chapter 2

## MIXTURES

### 2.1. Mechanics of Mixtures

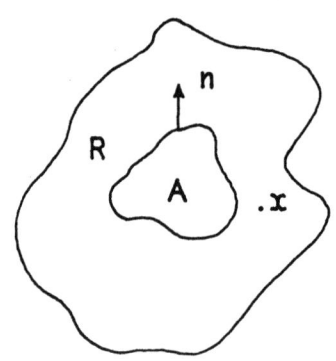

We consider a mixture of N constituents. For convenience, we identify the set of constituents with the set $\{1,2,...,N\}$. We reserve the letters $\alpha$ and $\beta$ for constituents. We will deal only with fluid mixtures; for such mixtures it is more convenient to work with the spatial description. Thus we let R denote a fixed region of space. (Fluid is allowed to flow into and out of R.)

<u>Terminology.</u>

| | | |
|---|---|---|
| $\rho_\alpha(x,t)$ | mass density of $\alpha$ ($\in R^+$) | |
| $v_\alpha(x,t)$ | velocity of $\alpha$ ($\in V$) | |
| $\xi_\alpha(x,t)$ | mass supply for $\alpha$ (due to chemical reactions)($\in R$) | |
| $\rho = \sum \rho_\alpha$ | total density | (2.1.1) |
| $c_\alpha = \rho_\alpha/\rho$ | concentration of $\alpha$ | (2.1.2) |
| $v = \sum c_\alpha v_\alpha$ | (mean) velocity of the mixture | (2.1.3) |
| $u_\alpha = v_\alpha - v$ | diffusion velocity of $\alpha$ | (2.1.4) |

By (2.1.1.) - (2.1.4),

# Balance of Mass

$$\sum c_\alpha = 1, \quad \sum \varrho_\alpha u_\alpha = \sum c_\alpha u_\alpha = 0. \quad (2.1.5)$$

<u>Notation</u>. Given a function $(x,t) \mapsto f(x,t)$ we write $f'$ for the spatial time derivative (with respect to $t$ holding $x$ fixed), $\operatorname{grad} f$ for the spatial gradient (with respect to $x$ holding $t$ fixed), $\operatorname{div} f$ for the spatial divergence. Let $(x,t) \mapsto \varphi(x,t)$ be a scalar field, and let $(x,t) \mapsto w(x,t)$ be a vector field. We define

$$\dot{\varphi} = \varphi' + v \cdot \operatorname{grad} \varphi, \quad D_\alpha \varphi = \varphi' + v_\alpha \cdot \operatorname{grad} \varphi$$
$$\dot{w} = w' + (\operatorname{grad} w)v, \quad D_\alpha w = w' + (\operatorname{grad} w)v_\alpha. \quad (2.1.6)$$

Thus a superposed dot denotes the material time derivative with respect to the mean motion of the mixture; $D_\alpha$ designates the material time derivative with respect to the motion of constituent $\alpha$. By (2.1.4) and (2.1.6),

$$\dot{\varphi} = D_\alpha \varphi - (\operatorname{grad} \varphi) \cdot u_\alpha, \quad \dot{w} = D_\alpha w - (\operatorname{grad} w) u_\alpha. \quad (2.1.7)$$

<u>Balance of Mass</u>:

$$\frac{d}{dt} \int_A \varrho_\alpha dv + \int_{\partial A} \varrho_\alpha v_\alpha \cdot n \, da = \int_A \varrho \xi_\alpha dv \quad (2.1.8)$$

for every regular subregion $A$ of $R$ and for every $t$. If the fields involved are sufficiently smooth, (2.1.8) is equivalent to

$$\varrho'_\alpha + \operatorname{div}(\varrho_\alpha v_\alpha) = \varrho \xi_\alpha, \quad (2.1.9)$$

or, alternatively, using (2.1.6) and (2.1.7),

(2.1.10) $$D_\alpha \varrho_\alpha + \varrho_\alpha \operatorname{div} v_\alpha = \varrho \xi_\alpha,$$

(2.1.11) $$\dot{\varrho}_\alpha + \varrho_\alpha \operatorname{div} v_\alpha + u_\alpha \operatorname{grad} \varrho_\alpha = \varrho \xi_\alpha,$$

(2.1.12) $$\dot{\varrho}_\alpha + \operatorname{div}(\varrho_\alpha u_\alpha) + \varrho_\alpha \operatorname{div} v = \varrho \xi_\alpha.$$

We now assume that mass is conserved for the mixture as a whole ; more precisely, we assume that

(2.1.13) $$\sum \xi_\alpha = 0.$$

Then, summing (2.1.12) over $\alpha$, we conclude, with the aid of (2.1.1) and (2.1.5), that

$$\dot{\varrho} + \varrho \operatorname{div} v = 0,$$

which is the usual law of mass balance in continuum mechanics.

Exercise. The quantity

$$h_\alpha = \varrho_\alpha u_\alpha$$

is called the relative mass flux for $\alpha$. Show that

(2.1.14) $$\varrho \dot{c}_\alpha = -\operatorname{div} h_\alpha + \varrho \xi_\alpha.$$

Exercise. Let $\varphi$ and $\varphi_\alpha$ be scalar fields, let $w$ and $w_\alpha$ be vector fields, and suppose that

$$\varrho \varphi = \sum \varrho_\alpha \varphi_\alpha, \quad \varrho w = \sum \varrho_\alpha w_\alpha.$$

# Balance of Momentum

Using (2.1.2) and (2.1.14) prove that

$$\varrho\dot{\varphi} = \sum[\varrho_\alpha D_\alpha \varphi_\alpha - \text{div}(\varrho_\alpha \varphi_\alpha u_\alpha) + \varrho \xi_\alpha \varphi_\alpha], \quad (2.1.15)$$

$$\varrho\dot{w} = \sum[\varrho_\alpha D_\alpha w_\alpha - \text{div}(\varrho_\alpha w_\alpha \otimes u_\alpha) + \varrho \xi_\alpha w_\alpha]. \quad (2.1.16)$$

**Terminology.**

| | | |
|---|---|---|
| $T_\alpha(x,t)$ | stress tensor for $\alpha$ | (EL) |
| $b_\alpha(x,t)$ | body force on $\alpha$ | (EV) |
| $\ell_\alpha(x,t)$ | momentum supply for $\alpha$ | (EV) |

**Balance of Momentum:**

$$\frac{d}{dt}\int_A \varrho_\alpha v_\alpha dv + \int_{\partial A}\varrho_\alpha v_\alpha(v_\alpha\cdot n)da =$$
$$= \int_{\partial A}T_\alpha n\, da + \int_A \varrho_\alpha(\ell_\alpha + b_\alpha)dv + \int_A \varrho \xi_\alpha v_\alpha dv \quad (2.1.17)$$

for every subregion $A$ of $R$ and for every $t$. Equivalently,

$$(\varrho_\alpha v_\alpha)' + \text{div}(\varrho_\alpha v_\alpha \otimes v_\alpha) = \text{div} T_\alpha + \varrho_\alpha(\ell_\alpha + b_\alpha) + \varrho \xi_\alpha v_\alpha, \quad (2.1.18)$$

or, by (2.1.6) – (2.1.8),

$$\varrho_\alpha D_\alpha v_\alpha = \text{div} T_\alpha + \varrho_\alpha(\ell_\alpha + b_\alpha), \quad (2.1.19)$$

$$\varrho_\alpha \dot{v}_\alpha + \varrho_\alpha(\text{grad}\, v_\alpha)u_\alpha = \text{div} T_\alpha + \varrho_\alpha(\ell_\alpha + b_\alpha). \quad (2.1.20)$$

**Exercise.** Derive (2.1.19) and (2.1.20).

By (2.1.5) and (2.1.16) with $w = v$, $w_\alpha = v_\alpha$,

$$(2.1.21) \quad \varrho \dot{v} = \sum [\varrho_\alpha D_\alpha v_\alpha + \varrho \xi_\alpha v_\alpha - \text{div}(\varrho_\alpha u_\alpha \otimes u_\alpha)]$$

We now assume that

$$\sum (\varrho_\alpha l_\alpha + \varrho \xi_\alpha v_\alpha) = 0,$$

or equivalently, by (2.1.4) and (2.1.13),

$$(2.1.22) \quad \sum (\varrho_\alpha l_\alpha + \varrho \xi_\alpha u_\alpha) = 0;$$

in view of (2.1.17) or (2.1.18), (2.1.22) expresses the requirement that momentum be conserved for the mixture as a whole. We define (the inner part of) the total stress $T$ and the total body force $b$ by the relations

$$(2.1.23) \quad T = \sum T_\alpha, \quad \varrho b = \sum \varrho_\alpha b_\alpha.$$

Then, if we sum (2.1.19) over all $\alpha$ and use (2.1.21) – (2.1.23), we arrive at the law of momentum balance for the mixture:

$$(2.1.24) \quad \varrho \dot{v} = \text{div}(T - \sum \varrho_\alpha u_\alpha \otimes u_\alpha) + \varrho b.$$

If we take $\psi = \sum c_\alpha u_\alpha^2, \psi_\alpha = u_\alpha^2$ in (2.1.15) and use (2.1.7), we arrive at

$$(2.1.25) \quad \varrho \sum \overline{c_\alpha u_\alpha^2} = \sum [\varrho_\alpha D_\alpha(u_\alpha^2) - \text{div}(\varrho_\alpha u_\alpha^2 u_\alpha) + \varrho \xi_\alpha u_\alpha^2].$$

Next, by (2.1.4), (2.1.5), and (2.1.7),

$$(2.1.26) \quad \sum \varrho_\alpha (D_\alpha v_\alpha) \cdot u_\alpha = \sum \left[ \frac{\varrho_\alpha}{2} D_\alpha(u_\alpha^2) + \varrho_\alpha (u_\alpha \otimes u_\alpha) \cdot \text{grad}\, v \right].$$

# The First Two Laws

Since
$$(\text{div}\, T_\alpha)\cdot u_\alpha = \text{div}(T_\alpha^T u_\alpha) - T_\alpha\cdot\text{grad}\, u_\alpha,$$
we conclude from (2.1.4), (2.1.19), (2.1.23), (2.1.25) and (2.1.26) that

$$\tfrac{1}{2}\varrho \overline{\sum c_\alpha u_\alpha^2}^{\cdot} = \text{div}\left[\sum\left(T_\alpha^T - \frac{\varrho_\alpha u_\alpha^2}{2}\right)u_\alpha\right] + (T - \sum \varrho_\alpha u_\alpha \otimes u_\alpha)\cdot\text{grad}\, v +$$
$$+ \sum\left[-T_\alpha\cdot\text{grad}\, v_\alpha + n_\alpha\cdot u_\alpha + \varrho_\alpha b_\alpha\cdot u_\alpha\right] \qquad (2.1.27)$$

where
$$n_\alpha = \varrho_\alpha \ell_\alpha + \tfrac{1}{2}\varrho \xi_\alpha u_\alpha. \qquad (2.1.28)$$

## 2.2. Thermodynamics of Mixtures

Terminology.

$\varepsilon(x,t)$     energy    $(\in R)$

$q(x,t)$     heat flux    $(\in V)$

$j(x,t)$     diffusive energy flux    $(\in V)$

$\gamma(x,t)$     heat supply    $(\in R)$

$\theta(x,t)$     temperature    $(\in R^+)$

$\eta(x,t)$     entropy    $(\in R)$

Balance of energy:

$$\varrho\overline{\left(\varepsilon + \frac{v^2}{2}\right)}^{\cdot} = -\text{div}(q+j) + \text{div}\left[(T^T - \sum \varrho_\alpha u_\alpha \otimes u_\alpha)v\right] + \sum \varrho_\alpha b_\alpha\cdot v_\alpha + \varrho\gamma.$$
$$(2.2.1)$$

Growth of entropy:

$$\varrho\dot\eta \geq -\text{div}\left(\frac{q}{\theta}\right) + \frac{\gamma}{\theta}. \qquad (2.2.2)$$

In view of (2.1.24), we can rewrite (2.2.1) as follows

$$\varrho \dot{\varepsilon} = -\text{div}(q+j) + (T - \sum \varrho_\alpha u_\alpha \otimes u_\alpha) \cdot \text{grad} \, v + \sum \varrho_\alpha b_\alpha \cdot u_\alpha + \varrho \gamma.$$
(2.2.3)

We now define

(2.2.4) $$\varepsilon_I = \varepsilon - \frac{1}{2} \sum c_\alpha u_\alpha^2 ,$$

(2.2.5) $$k = j + \sum \left(T_\alpha^T - \frac{1}{2} \varrho_\alpha u_\alpha^2 1\right) u_\alpha ;$$

$e_I$ is the inner part of the internal energy. If we use (2.1.27) to eliminate the body force term in (2.2.3), we find, with the aid of (2.2.4) and (2.2.5), that

(2.2.6) $$\varrho \dot{\varepsilon}_I = -\text{div}(q+k) + \sum T_\alpha \cdot \text{grad} \, v_\alpha - \sum n_\alpha \cdot u_\alpha + \varrho \gamma .$$

We define (the inner part of) the <u>free energy</u> by

(2.2.7) $$\psi = \varepsilon_I - \theta \eta .$$

Then (2.2.2) and (2.2.6) yield the reduced dissipation inequality:

$$\varrho(\dot{\psi} + \eta \dot{\theta}) - \sum T_\alpha \cdot \text{grad} \, v_\alpha + \text{div} \, k + \sum n_\alpha \cdot u_\alpha + \frac{1}{\theta} q \cdot g \leq 0 .$$
(2.2.8)

<u>Notation</u>. Given any quantity $f_\alpha$ associated with the mixture, we write

(2.2.9) $$\vec{f} = (f_1, \ldots, f_N)$$

We now list the basic equations in the form that we will use them:

$$\varrho'_\alpha + \text{div}(\varrho_\alpha v_\alpha) = \varrho \xi_\alpha , \qquad (2.1.9)$$

$$\varrho_\alpha D_\alpha v_\alpha = \text{div} T_\alpha + \varrho_\alpha (\ell_\alpha + b_\alpha) , \qquad (2.1.19)$$

$$\varrho \dot{\varepsilon} = -\text{div}(q+j) + (T - \sum \varrho_\alpha u_\alpha \otimes u_\alpha) \cdot \text{grad} v + \sum \varrho_\alpha b_\alpha \cdot u_\alpha + \varrho \gamma , \qquad (2.2.3)$$

$$\varrho(\dot{\Psi} + \eta \dot{\theta}) - \sum T_\alpha \cdot \text{grad} v_\alpha + \text{div} k + \sum n_\alpha \cdot u_\alpha + \frac{1}{\theta} q \cdot g \leq 0 , \qquad (2.2.8)$$

with

$$\sum \xi_\alpha = 0 , \qquad (2.1.13)$$

$$\sum (\varrho_\alpha \ell_\alpha + \varrho \xi_\alpha u_\alpha) = 0 , \qquad (2.1.22)$$

and where

$$\Psi = \varepsilon_I - \theta \eta = \varepsilon - \frac{1}{2} \sum c_\alpha u_\alpha^2 - \theta \eta , (2.2.4) \; (2.2.7)$$

$$k = j + \sum \left( T_\alpha^T - \frac{1}{2} \varrho_\alpha u_\alpha^2 1 \right) u_\alpha . \qquad (2.2.5)$$

<u>Definition.</u> Let $A$ be a regular subregion of $R$, and let $T$ be a time interval. A <u>process</u> in $A$ during $T$ is an ordered array $p = [\vec{\varrho}, \theta, \vec{v}, \Psi, \vec{T}, \eta, q, j, \vec{\ell}, \vec{\xi}]$ such that

$$\left. \begin{array}{l} \vec{\varrho} : A \times T \to (R^+)^N \\ \\ \theta : A \times T \to R^+ \end{array} \right\} \underline{\text{are class}} \; C^2 ,$$

$$\left.\begin{array}{l} \vec{v} : A\times T \to V^N \\ \Psi : A\times T \to R \\ \vec{T} : A\times T \to L^N \\ \eta : A\times T \to R \\ q : A\times T \to V \\ \dot{\downarrow} : A\times T \to V \end{array}\right\} \text{ are class } C^1,$$

$$\left.\begin{array}{l} \vec{\ell} : A\times T \to V^N \\ \vec{\xi} : A\times T \to R^N \end{array}\right\} \text{ are continuous .}$$

Remark. Given a process $p$, (2.1.19) and (2.2.3) can be used to compute the body forces $\vec{b}(p)$ and the heat supply $\gamma(p)$ necessary to support the process.

## 2.3. Constitutive Relations

We consider a mixture defined by constitutive relations in which

(2.3.1)
$$(\Psi, \vec{T}, \eta, q, \dot{\downarrow}, \vec{\ell}, \vec{\xi})$$
are functions of
$$(\vec{\varrho}, \theta, \vec{f}, g, \vec{v}),$$

where
$$\vec{f} = \text{grad}\,\vec{\varrho}, \quad g = \text{grad}\,\theta.$$

Since we are dealing with fluid mixtures, we assume that the response functions $\hat{\Psi}, \hat{T}_\alpha$, etc. are isotropic. In addition, we assume that the response functions are of class $C^3$ on their domain

$$D = (R^+)^N \times R^+ \times V^N \times V \times V^N,$$

and that $\hat{t}_\alpha$ and $\hat{\xi}_\alpha$ are consistent with (2.1.13) and (2.1.22). Further, to make our theory consistent with the principle od material frame indifference, we require that

$$\hat{\varphi}(\vec{\varrho},\theta,\vec{f},g,\vec{v}) = \hat{\varphi}(\vec{\varrho},\theta,\vec{f},g,\vec{v}+\vec{a}), \quad (2.3.2)$$

whenever $\hat{\varphi}$ is one of the response functions in (2.3.1) and

$$\vec{a} = (a,a,\ldots,a). \quad (2.3.3)$$

By (2.3.2),

$$\sum \frac{\partial \hat{\varphi}}{\partial v_\alpha} = 0 \quad (2.3.4)$$

and

$$\hat{\varphi}(\vec{\varrho},\theta,\vec{f},g,\vec{v}) = \hat{\varphi}(\vec{\varrho},\theta,\vec{f},g,\vec{u}), \quad (2.3.5)$$

where

$$\vec{u} = (u_1,\ldots,u_N)$$

is the vector $(\in V^N)$ of diffusion velocities.

Assumption. Given:

(i) a closed ball $A_0 \subset R$ and a time interval $T_0 = [0, \tau_0]$,

(ii) an analytic initial density distribution

$$\vec{\varrho}_0 : A_0 \to (R^+)^N ;$$

(iii) an analytic velocity field

$$\vec{v} : A_0 \times T_0 \to V^N ;$$

(iv) an analytic temperature field

$$\theta : A_0 \times T_0 \to R^+ ;$$

there exists a closed ball $A \subset A_0$, a time interval $T = [0, \tau] \subset T_0$, and a class $C^2$ solution

$$\vec{\varrho} : A \times [0, \tau] \to (R^+)^N$$

of balance of mass (2.1.9), i.e. of

$$\varrho'_\alpha + \text{div}(\varrho_\alpha v_\alpha) = \varrho \hat{\xi}_\alpha(\vec{\varrho}, \theta, \vec{f}, g, \vec{v})$$

on $A \times [0, \tau]$, such that

$$\vec{\varrho}(x, 0) = \vec{\varrho}_0(x) \quad (\forall x \in A)$$

Definition. A process p is called a <u>constitutive process</u> if it is a consistent with the constitutive equations (2.3.1) and balance of mass (2.1.9).

Remark. The assumption given above insures that given the data specified in (i) – (iv), there exists an associated constitutive process.

Note that, by (2.3.1) and (2.1.28),

$$\eta = \hat{\eta}(\vec{\varrho}, \theta, \vec{f}, g, \vec{v}), \quad k = \hat{k}(\vec{\varrho}, \theta, \vec{f}, g, \vec{v}). \quad (2.3.6)$$

Notation.

$$div_0 \hat{k} = \sum \frac{\partial \hat{k}}{\partial \varrho_\alpha} \cdot grad\, \varrho + \frac{\partial \hat{k}}{\partial \theta} \cdot grad\, \theta, \quad (2.3.7)$$

$$div_1 \hat{k} = \sum \frac{\partial \hat{k}}{\partial f_\alpha} \cdot grad^2 \varrho_\alpha + \frac{\partial \hat{k}}{\partial g} \cdot grad^2 \theta.$$

Thus $\quad div\, k = div_0 \hat{k} + div_1 \hat{k} + \sum \left(\frac{\partial \hat{k}}{\partial v_\alpha}\right)^T \cdot grad\, v_\alpha. \quad (2.3.8)$

## 2.4. Consequences of the Second Law

Theorem. <u>A necessary and sufficient condition that every constitutive process obey the reduced dissipation inequality (2.2.8) is that the following four statements be true</u>

(i) <u>The total stress is a pressure</u> :

$$T = -p1. \quad (2.4.1.)$$

(ii) <u>Ψ, p and η are independent of</u> $\vec{f} = grad\,\vec{\varrho}, g = grad\,\theta$ <u>and</u> $\vec{v}$ :

$$\Psi = \hat{\Psi}(\vec{\varrho}, \theta), \quad p = \hat{p}(\vec{\varrho}, \theta), \quad \eta = \hat{\eta}(\vec{\varrho}, \theta) ; \quad (2.4.2)$$

in addition,

$$\hat{p} = \varrho \sum \varrho_\alpha \frac{\partial \hat{\Psi}}{\partial \varrho_\alpha}, \quad \hat{\eta} = -\frac{\partial \hat{\Psi}}{\partial \theta}. \quad (2.4.3)$$

(iii) <u>The constituent stress $\hat{T}_\alpha$ is given by</u>

(2.4.4.) $$\hat{T}_\alpha = \left(\frac{\partial \hat{k}}{\partial v_\alpha}\right)^T - \varrho \varrho_\alpha \frac{\partial \hat{\psi}}{\partial \varrho_\alpha} \mathbf{1}.$$

(iv) <u>In every constitutive process</u>

$$div_0\hat{k} + \varrho\sum \frac{\partial \hat{\psi}}{\partial \varrho_\alpha}[\varrho\hat{\xi}_\alpha - u_\alpha\cdot grad\,\varrho_\alpha] + \sum \hat{n}_\alpha \cdot u_\alpha + \frac{1}{\theta}\hat{q}\cdot g \leq 0,$$

(2.4.5) $$div_1\hat{k} = 0.$$

Proof. Sufficiency follows immediately upon direct substitution. To establish the necessity of (i) – (iv) we assume that every constitutive process obeys (2.2.8); thus, using (2.1.12) (recall that every constitutive process automatically obeys balance of mass) and (2.3.8):

$$\varrho\left(\frac{\partial \hat{\psi}}{\partial \theta} + \hat{\eta}\right)\dot{\theta} - \sum\left[\varrho\varrho_\alpha\frac{\partial \hat{\psi}}{\partial \varrho_\alpha}\mathbf{1} + \hat{T}_\alpha - \left(\frac{\partial \hat{k}}{\partial v_\alpha}\right)^T\right]\cdot grad\,v_\alpha +$$

$$+ \varrho\left[\sum\frac{\partial \hat{\psi}}{\partial f_\alpha}\cdot \dot{f}_\alpha + \frac{\partial \hat{\psi}}{\partial g}\cdot \dot{g} + \sum\frac{\partial \hat{\psi}}{\partial v_\alpha}\cdot \dot{v}_\alpha\right] + \varrho\sum\frac{\partial \hat{\psi}}{\partial \varrho_\alpha}[\varrho\hat{\xi}_\alpha - u_\alpha\cdot grad\,\varrho_\alpha] +$$

(2.4.6) $$+ div_0\hat{k} + div_1\hat{k} + \sum\hat{n}_\alpha\cdot u_\alpha + \frac{1}{\theta}\hat{q}\,g \leq 0.$$

It follows in a manner quite similar to that used to establish the main theorem in Section 2 that $\dot{\theta}$, $\dot{f}_\alpha$, $\dot{g}$, $\dot{v}_\alpha$ and $grad\,v_\alpha$ can be arbitrarily specified in (2.4.6), and this observation leads to $(2.4.2)_{1,3}$, $(2.4.3)_2$, and (2.4.4). Next, it follows from (2.3.4) with, $\hat{\varphi} = \hat{k}$, (2.1.23), (2.4.4) that (2.4.1) and $(2.4.2)_2$ hold. Next, $div_1\hat{k}$ is the only term in (2.4.6) involving second gradients; thus, by (2.3.7)

(2.4.7) $$sym\left(\frac{\partial \hat{k}}{\partial f_\alpha}\right) = sym\left(\frac{\partial \hat{k}}{\partial g}\right) = 0,$$

# The Gibbs' Relation

which implies $(2.4.5)_2$. Finally, $(2.4.5)$ follows from the results established thus far.

We now assume that (i) – (iv) of the above theorem hold. Let

$$\nu = \hat{\nu}(\vec{\varrho}, \theta)$$

be any given function; we define the <u>chemical potential</u> of constituent $\alpha$ by the relation

$$\mu_\alpha = \hat{\mu}_\alpha(\vec{\varrho}, \theta) = \varrho \frac{\partial \hat{\psi}(\vec{\varrho}, \theta)}{\partial \varrho_\alpha} + \hat{\nu}(\vec{\varrho}, \theta) . \qquad (2.4.8)$$

Thus the functions $\hat{\mu}_\alpha$ depend on the choice of $\hat{\nu}$; however, the difference between the potentials of any two constituents is independent of $\hat{\nu}$, and, as we shall see, all of our results will be independent of the choice of $\hat{\nu}$.

<u>Theorem.</u> <u>In each constitutive process we have the Gibbs' relation</u>

$$\dot{\psi} = \frac{1}{\varrho^2} p \dot{\varrho} - \eta \dot{\theta} + \sum \mu_\alpha \dot{c}_\alpha .$$

<u>Exercise.</u> Prove this theorem. Hint: Use the fact that, by $(2.1.5)$,

$$\sum \dot{c}_\alpha = 0 .$$

<u>Lemma.</u> Let $f: V^M \to V$ <u>be an isotropic function</u>, i.e.

$$f(Qw_1, Qw_2, \ldots, Qw_M) = Qf(w_1, w_2, \ldots, w_M) \qquad (2.4.9)$$

for every $w = (w_1, w_2, \ldots, w_M) \in V^M$ <u>and for every orthogonal tensor</u>

Q. Further, assume that $f$ is class $C^3$ and that for every $w \in V^M$ and $m \in \{1,2,\ldots,M\}$ the tensor $\dfrac{\partial f(w)}{\partial w_m}$ is skew. Then $f \equiv 0$.

The proof of this lemma will be given at the end of the section.

<u>Theorem.</u> The diffusive energy flux $\vec{j} = \hat{\vec{j}}(\vec{g},\theta,\vec{f},g,\vec{v})$ <u>vanishes when the diffusive velocities vanish.</u> In fact,
$$\vec{j} = \sum \mu_\alpha \varrho_\alpha u_\alpha + O(|\vec{u}|^2) \quad \text{as} \quad \vec{u} \to \vec{0}$$
<u>Proof.</u> Let $\Omega = (\vec{g},\theta,\vec{f},g)$. Since $k = \hat{k}(\Omega,\vec{v})$ is an isotropic function, it follows from (2.4.7) and the Lemma that
$$\hat{k}(\Omega,\vec{0}) = 0,$$
and we conclude from (2.2.5) that
$$\hat{\vec{j}}(\Omega,\vec{0}) = 0.$$
Next, by (2.1.3) and (2.1.4),
$$u_\beta = v_\beta - \sum c_\alpha v_\alpha ,$$
so that

(2.4.10) $$\frac{\partial u_\beta}{\partial v_\alpha} = (\delta_{\alpha\beta} - c_\alpha)\mathbf{1}.$$

Thus, if we differentiate (2.2.5) with respect to $v_\alpha$ and use (2.1.23), (2.4.1), and (2.4.4), we arrive at
$$\frac{\partial \hat{\vec{j}}}{\partial v_\alpha} = \left(\varrho\varrho_\alpha \frac{\partial \psi}{\partial \varrho_\alpha} - c_\alpha p\right)\mathbf{1} \quad \text{when} \quad \vec{v} = \vec{0}.$$

Therefore,

$$\hat{j}(\Omega,\vec{v}) = \sum\left(\varrho\varrho_\alpha\frac{\partial\Psi}{\partial\varrho_\alpha} - c_\alpha p\right)v_\alpha + O(|\vec{v}|^2) \quad \text{as} \quad \vec{v}\to\vec{0}.$$

But by (2.3.5) $\hat{j}(\Omega,\vec{v}) = \hat{j}(\Omega,\vec{u})$; thus the desired result follows from (2.1.5) and (2.4.8).

Let **n** be a unit vector. The quantity

$$t_\alpha(n) = T_\alpha n$$

is the stress vector for constituent $\alpha$ (corresponding to **n**) By (2.3.1), $t_\alpha(n)$ is a function of $(\vec{\varrho},\theta,\vec{f},g,\vec{v})$. We call the matrix

$$\left\|\frac{\partial t_\alpha(n)}{\partial v_\beta}\right\|$$

(with tensor entries) the stress-diffusion matrix.

**Theorem.** The stress-diffusion matrix is symmetric, i.e.

$$\frac{\partial t_\alpha(n)}{\partial v_\beta} = \left(\frac{\partial t_\beta(n)}{\partial v_\alpha}\right)^T.$$

Exercise. Prove this theorem. Hint: Use (2.4.4).

Proof of the Lemma. Choose an orthonormal basis for **V** and let $f<i>$ and $w_m<i>$ ($i = 1,2,3$) denote the corresponding components of **f** and $w_m$. By hypothesis,

$$\frac{\partial f<i>}{\partial w_m<j>} = -\frac{\partial f<j>}{\partial w_m<i>}, \qquad (2.4.11)$$

so that

$$(2.4.12) \qquad \frac{\partial f_{\langle i \rangle}}{\partial w_{m\langle i \rangle}} = 0.$$

Equations (2.4.11) and (2.4.12) imply that

$$(2.4.13) \qquad \frac{\partial^2 f_{\langle i \rangle}}{\partial w_{m\langle j \rangle} \partial w_{n\langle j \rangle}} = -\frac{\partial^2 f_{\langle j \rangle}}{\partial w_{m\langle j \rangle} \partial w_{n\langle i \rangle}} = 0 \, ;$$

hence

$$(2.4.14) \qquad \frac{\partial^3 f_{\langle i \rangle}}{\partial w_{m\langle j \rangle} \partial w_{n\langle k \rangle} \partial w_{p\langle \ell \rangle}} = 0,$$

since two of $i, j, k$ and $\ell$ must coincide. Therefore $f$ must have the form

$$(2.4.15) \qquad f(w) = f(0) + F(w) + G(w,w),$$

where $F: V^M \to V$ is linear and $G: V^M \times V^M \to V$ is symmetric and bilinear. It follows from (2.4.15) and (2.4.10) with $Q = -1$ that $f(0) = 0$ and $G = 0$; therefore

$$(2.4.16) \qquad f(w) = F(w) = \sum_{m=1}^{M} F_m w_m,$$

where each $F_m$ is a tensor. Further,

$$(2.4.17) \qquad F_m = \partial f(w)/\partial w_m,$$

so that $F_m$ is skew. Finally, by (2.4.10) and (2.4.16),

$$(2.4.18) \qquad QF_m = F_m Q$$

for every orthogonal $Q$, and the only skew tensor with this property is $F_m = 0$.

## 2.5. Results near Equilibrium

Let $(\vec{g}_0, \theta_0)$ be given, and for convenience, let

$$s = (\vec{g}, \theta, \vec{f}, g, \vec{v}), \quad s_0 = (\vec{g}_0, \theta_0, \vec{0}, 0, \vec{0}).$$

We call $s_0$ an <u>equilibrium state</u> provided

$$\hat{\xi}_\alpha(s_0) = 0 \qquad (2.5.1)$$

for every constituent $\alpha$. Let $h$ denote the left-hand side of (2.4.5). It then follows that

$$h = \hat{h}(s) = \sum \frac{\partial \hat{k}}{\partial \varrho_\alpha} \cdot f_\alpha + \frac{\partial \hat{k}}{\partial \theta} \cdot g + \sum \left( \hat{n}_\alpha - \varrho \frac{\partial \hat{\psi}}{\partial \varrho_\alpha} f_\alpha \right) \cdot u_\alpha + \frac{1}{\theta} \hat{q} \cdot g + \varrho \sum \hat{\mu}_\alpha \hat{\xi}_\alpha , \qquad (2.5.2)$$

$$\hat{h}(s) \leq 0 \qquad (2.5.3)$$

for every $s$ in the domain of the response functions. Here $\mu_\alpha$ is given by (2.4.8), and we have chosen the function $\hat{v}$ such that

$$\sum \mu_\alpha = 0. \qquad (2.5.4)$$

If $s_0$ is an equilibrium state, then (2.5.1) and (2.5.2) imply that $\hat{h}(s_0) = 0$, so that (2.5.5)
$\hat{h}$ <u>is a maximum at</u> $s = s_0$.
Thus

$$(2.5.6) \qquad \left(\frac{\partial \hat{h}}{\partial \varrho_\alpha}\right)_0 = \left(\frac{\partial \hat{h}}{\partial \theta}\right)_0 = 0,$$

where the subscript "0" indicates that the corresponding function is to be evaluated at $s = s_0$, and we have the following result:

$$(2.5.7) \qquad \sum_\alpha \overset{\circ}{\mu}_\alpha \left(\frac{\partial \hat{h}}{\partial \varrho_\beta}\right)_0 = \sum_\alpha \overset{\circ}{\mu}_\alpha \left(\frac{\partial \hat{\xi}_\alpha}{\partial \theta}\right)_0 = 0.$$

Here $\overset{\circ}{\mu}_\alpha = \hat{\mu}_\alpha(\vec{\varrho}_0, \theta_0)$. In addition, since $\hat{\xi}_\alpha$ is an isotropic function,

$$(2.5.8) \qquad \left(\frac{\partial \hat{\xi}_\alpha}{\partial f_\beta}\right)_0 = \left(\frac{\partial \hat{\xi}_\alpha}{\partial g}\right)_0 = \left(\frac{\partial \hat{\xi}_\alpha}{\partial v_\beta}\right)_0 = 0.$$

It follows from (2.5.1), (2.5.7), and (2.5.8) that

$$(2.5.9) \qquad \sum_\alpha \hat{\xi}_\alpha(s)\overset{\circ}{\mu}_\alpha = O(|s - s_0|^2) \quad \text{as} \quad s \to s_0,$$

where

$$(2.5.10) \qquad |s - s_0| = |\vec{\varrho} - \vec{\varrho}_0| + |\theta - \theta_0| + |\vec{f}| + |g| + |\vec{v}|.$$

We say that an equilibrium state $s_0$ is strong provided

$$(2.5.11) \qquad \sum_\alpha \hat{\xi}_\alpha(s)\overset{\circ}{\mu}_\alpha = O(|s - s_0|^3) \quad \text{as} \quad s \to s_0.$$

Remark. To see that this is a natural generalization of the usual notion of strong equilibrium, we assume, for the time being, that there are R independent chemical reactions:

$$\hat{\xi}_\alpha = \sum_{r=1}^{R} \nu_{\alpha r} J_r \; , \qquad (2.5.12)$$

where $\nu_{\alpha r}$ is the stoichiometric coefficient of constituent $\alpha$ in the reaction $r$ divided by the molecular mass of $\alpha$, and $J_r = J_r(s)$ is the reaction rate of reaction $r$. The chemical affinity of reaction $r$ is defined by

$$A_r = A_r(\vec{\mathbf{g}},\theta) = \sum_\alpha \nu_{\alpha r} \mu_\alpha(\vec{\mathbf{g}},\theta) \; . \qquad (2.5.13)$$

In this instance it is customary to call $s_0$ a "strong equilibrium state" provided

$$J_r(s_0) = A_r(\vec{\mathbf{g}}_0,\theta_0) = 0 \; . \qquad (2.5.14)$$

In view of (2.5.12), the first of (2.5.14) implies (2.5.1). Further, by (2.5.12) - (2.5.14),

$$\sum \hat{\xi}_\alpha(s) \overset{\circ}{\mu}_\alpha \equiv 0 \; . \qquad (2.5.15)$$

Thus our notion of a strong equilibrium state is somewhat weaker than the standard definition. For all of our results it suffices to use the definition containing (2.5.11).

With a view toward determining the behaviour of the response functions near equilibrium, we first determine some of the more obvious consequences of isotropy. First of all there exist scalar functions $p_\alpha(\vec{\mathbf{g}},\theta)$ such that

$$(2.5.16) \qquad \hat{T}_\alpha(\vec{\varrho},\theta,\vec{0},0,\vec{0}) = -\hat{p}_\alpha(\vec{\varrho},\theta)1 .$$

Next, if $s_0$ is an equilibrium state, then, clearly,

$$\hat{z}(s_0) = \left(\frac{\partial z}{\partial \varrho_\alpha}\right)_0 = \left(\frac{\partial \hat{z}}{\partial \theta}\right)_0 = 0 \quad \text{for} \quad \hat{z} = \hat{q} \quad \text{or} \quad \hat{\ell}_\beta .$$
(2.5.17)

Further, there exist scalars $\varkappa, \varkappa_\alpha, \gamma_{\alpha\beta}, \gamma_\alpha$ and $\lambda_{\alpha\beta}$ such that

$$(2.5.18) \quad \begin{aligned} \left(\frac{\partial \hat{q}}{\partial g}\right)_0 &= -\varkappa 1 , & \left(\frac{\partial \hat{q}}{\partial v_\alpha}\right)_0 &= -\varkappa_\alpha 1 , \\ \left(\varrho_\alpha \frac{\hat{\ell}_\alpha}{\partial v_\beta}\right)_0 &= -\gamma_{\alpha\beta} 1 , & \left(\varrho_\alpha \frac{\partial \hat{\ell}_\alpha}{\partial g}\right)_0 &= -\gamma_\alpha 1 , \\ \left(\frac{1}{\varrho_\alpha}\frac{\partial \hat{p}_\alpha}{\partial \varrho_\beta}\right)_0 1 - \left(\frac{\partial \hat{\ell}_\alpha}{\partial f_\beta}\right)_0 &= \lambda_{\alpha\beta} 1 , \end{aligned}$$

and, in view of (2.1.13), (2.1.22), (2.3.4), (2.4.10), and (2.5.1),

$$(2.5.19) \quad \sum \varkappa_\alpha = \sum_\alpha \gamma_{\alpha\beta} = \sum_\beta \gamma_{\alpha\beta} = \sum \gamma_\alpha = 0 .$$

We call $\varkappa$ the <u>conductivity</u> and $\|\gamma_{\alpha\beta}\|$ the <u>momentum supply matrix</u>. The matrix $\|\lambda_{\alpha\beta}\|$ is of importance in applications. Indeed if we consider the <u>linearized</u> system of momentum equations appropriate for small departures from the equilibrium state $s_0$, then (after dividing by $\varrho_\alpha$) the term involving $\text{grad}\,\varrho_\alpha$ in the $\alpha$-th equations has the form $\sum_\beta \lambda_{\alpha\beta}\text{grad}\,\varrho_\beta$. For this reason we call $\|\lambda_{\alpha\beta}\|$ the __elasticity matrix__.

__Theorem.__ _Let $s_0$ be a strong equilibrium state._
_Then_

$$\lambda_{\beta\alpha} = \left(\frac{\partial^2(\varrho\hat{\psi})}{\partial\varrho_\alpha\partial\varrho_\beta}\right)_0, \qquad (2.5.20)$$

$$\left(\frac{\partial q}{\partial f_\alpha}\right)_0 = 0, \qquad (2.5.21)$$

$$\frac{\varkappa}{\theta_0}a^2 + \sum_\beta\left[\frac{\varkappa_\beta}{\theta_0} + \gamma_\beta + \left(\frac{\partial\hat{p}_\beta}{\partial\theta}\right)_0 + \left(\varrho\varrho_\beta\frac{\partial\hat{\eta}}{\partial\varrho_\beta}\right)_0\right]a\omega_\beta +$$

$$+ \sum_{\alpha,\beta}\gamma_{\alpha\beta}\omega_\alpha\omega_\beta \geq 0 \quad \text{for all} \quad a,\omega_1,\ldots,\omega_N \in R;$$
$$(2.5.22)$$

_so that, in particular, the conductivity $\varkappa \geq 0$ and the momentum supply matrix_ $\|\gamma_{\alpha\beta}\|$ _is positive semi-definite._

__Proof.__ By (2.5.4) and (2.5.11),

$$\left(\frac{\partial^2(\sum\hat{\mu}_\alpha\hat{\xi}_\alpha)}{\partial a \partial d}\right)_0 = 0 \quad \text{whenever} \quad a = f_\beta, g, \quad \text{or} \quad v_\beta \quad \text{and}$$

$$d = f_\gamma, g, \quad \text{or} \quad v_\gamma.$$

Next, since

$$\hat{k}(\vec{\varrho},\theta,\vec{f},g,\vec{0}) = 0, \qquad (2.5.23)$$

it follows from (2.5.3) and (2.5.11) that

$$\left(\frac{\partial^2 \hat{h}}{\partial f_\alpha \partial f_\beta}\right)_0 = 0. \qquad (2.5.24)$$

Thus, in view of (5),

$$\left(\frac{\partial^2 \hat{h}}{\partial f_\alpha \partial g}\right)_0 = \left(\frac{\partial^2 \hat{h}}{\partial f_\alpha \partial v_\beta}\right)_0 = 0. \qquad (2.5.25)$$

The first of (2.5.25), in conjunction (2.5.2), (2.5.11), and (2.5.23), yields (2.5.21). On the other hand, the second of (2.5.25), (2.5.2), and (2.5.11) imply

$$\left(\frac{\partial^2 \hat{k}}{\partial \varrho_\alpha \partial v_\beta}\right)_0^T + \sum_\gamma \left(\frac{\partial u_\gamma}{\partial v_\beta}\right)_0^T \left(\frac{\partial \hat{n}_\gamma}{\partial f_\alpha}\right)_0 - \left(\varrho \frac{\partial \hat{\psi}}{\partial \varrho_\alpha}\right)_0 \left(\frac{\partial u_\alpha}{\partial v_\beta}\right)_0^T = 0.$$
(2.5.26)

By (2.1.22), (2.1.28), and (2.5.1),

$$\left(\frac{\partial \hat{n}_\alpha}{\partial a}\right)_0 = \left(\varrho_\alpha \frac{\partial \hat{\ell}_\alpha}{\partial a}\right)_0 \quad \text{and} \quad \sum \left(\frac{\partial \hat{n}_\alpha}{\partial a}\right)_0 = 0 \quad \text{whenever}$$
(2.5.27)
$$a = f_\beta, g, \quad \text{or} \quad v_\beta,$$

and (2.4.4), (2.4.10), (2.5.16), (2.5.18), (2.5.26) and (2.5.27) yield, after some manipulation, the result (2.5.20)

Next, by (2.5.5).

$$a \cdot \left(\frac{\partial^2 \hat{h}}{\partial g^2}\right)_0 a + \sum_\beta a \cdot \left(\frac{\partial^2 \hat{h}}{\partial v_\beta \partial g}\right)_0 w_\beta + \sum_{\alpha,\beta} w_\alpha \cdot \left(\frac{\partial^2 \hat{h}}{\partial v_\beta \partial v_\alpha}\right)_0 w_\beta \leq 0 \quad (2.5.28)$$

for all vectors $a, w_1, \ldots, w_N$. If we take $a = e$ and $w_\alpha = \omega_\alpha e$ in (2.3.28), where $e$ is a unit vector, and use (2.4.3), (2.4.4), (2.4.10), (2.5.1), (2.5.11), (2.5.16), (2.5.18), (2.5.23) and (2.5.27), we are led to (2.5.22).

As a direct consequence of (2.5.20) we have the following important result.

Corollary. Let $s_0$ be a strong equilibrium state. Then the elasticity matrix is symmetric:

$$\lambda_{\alpha\beta} = \lambda_{\beta\alpha}. \quad (2.5.29)$$

The next corollary follows from (2.5.4), (2.5.27) − (2.5.19) and (2.5.21); it asserts that near a strong equilibrium state to within terms of $O(|s-s_0|^2)$ $\underset{\sim}{q}$ depends linearly on $\mathbf{grad}\,\theta$ and $\vec{u}$.

Corollary. Let $s_0$ be a strong equilibrium state. Then

$$q = -\varkappa g - \sum \varkappa_\alpha u_\alpha + O(|s-s_0|^2) \quad \text{as} \quad s \to s_0, \quad (2.5.30)$$

where $q = \hat{q}(s)$.

In view of (2.5.2), we can take $\varrho, c_1, \ldots, c_N$ as independent variables in place of $\varrho_1, \ldots, \varrho_N$, i.e., e.g.,

$$(2.5.31) \quad \hat{\mu}_\alpha(\varrho_1,\ldots,\varrho_N,\theta) = \hat{\mu}_\alpha(c_1\varrho,\ldots,c_N\varrho,\theta) = \tilde{\mu}_\alpha(\varrho,\theta,c_1,\ldots,c_N).$$

By $(2.5.5)_1$ and (2.5.4), the vectors $\vec{c} = (c_1,\ldots,c_N)$ and $\vec{\mu} = (\mu_1,\ldots,\mu_N)$ both lie on planes in $R^N$ of dimension $N-1$.
We assume that the mapping

$$(2.5.32) \quad \vec{c} \longrightarrow \vec{\tilde{\mu}}(\varrho,\theta,\vec{c})$$

is invertible in some neighborhood of $s_0$. Then in this neighborhood we can express the mass supply as follows:

$$(2.5.33) \quad \xi_\alpha = \tilde{\xi}_\alpha(\varrho,\theta,\vec{\mu},\vec{f},g,\vec{v}).$$

Let

$$(2.5.34)^{(*)} \quad \tau_{\alpha\beta} = -\left(\frac{\partial \tilde{\xi}_\alpha}{\partial \mu_\beta}\right)_0 ;$$

---

(*) The derivative $\left(\dfrac{\partial \tilde{\xi}_\alpha}{\partial \mu_1}, \ldots, \dfrac{\partial \tilde{\xi}_\alpha}{\partial \mu_N}\right)$ lies in the tangent space $\{(\varphi_1,\ldots,\varphi_N) \in R^N \mid \sum \varphi_\alpha = 0\}$.

we call $\|\tau_{\alpha\beta}\|$ the mass supply matrix. The next theorem shows that to within terms of $O(|s - s_0|^2)$ $\tilde{\xi}_\alpha$ depends only on $\mu_1, \ldots, \mu_N$.

Theorem. <u>Let $s_0$ be a strong equilibrium state and assume that the mapping (2.5.32) is invertible in some neighborhood of $s_0$.</u>
Then

$$\tilde{\xi}_\alpha = -\sum_\beta \tau_{\alpha\beta}(\mu_\beta - \overset{\circ}{\mu}_\beta) + O(|s - s_0|^2), \qquad (2.5.35)$$

where $\tilde{\xi}_\alpha = \hat{\tilde{\xi}}_\alpha(s)$. <u>Moreover, the mass supply matrix $\|\tau_{\alpha\beta}\|$ is positive semi-definite and</u>

$$\sum_\alpha \overset{\circ}{\mu}_\alpha \tau_{\alpha\beta} = 0. \qquad (2.5.36)$$

Proof. First, since $\tilde{\xi}_\alpha$ is isotropic,

$$\left(\frac{\partial \tilde{\xi}_\alpha}{\partial f_\beta}\right)_0 = \left(\frac{\partial \tilde{\xi}_\alpha}{\partial g}\right)_0 = \left(\frac{\partial \tilde{\xi}_\alpha}{\partial v_\beta}\right)_0 = 0. \qquad (2.5.37)$$

if

$$\vec{f} = \vec{0}, \quad g = 0, \quad \vec{v} = \vec{0}, \qquad (2.5.38)$$

then, letting

$$\mu'_\alpha = \mu_\alpha - \overset{\circ}{\mu}_\alpha, \quad \varrho' = \varrho - \varrho_0, \quad \theta' = \theta - \theta_0,$$
$$\varepsilon = \sum|\mu'_\alpha| + |\varrho'| + |\theta'|, \qquad (2.5.39)$$

we conclude from (2.5.1), (2.5.2), (2.5.11), (2.5.33) and (2.5.34) that

$$0 \geq \sum \tilde{\xi}_\alpha \mu_\alpha = \sum \tilde{\xi}_\alpha \mu'_\alpha + O(\varepsilon^3) = \sum_{\alpha,\beta}\left[-\tau_{\alpha\beta}\mu'_\beta + \left(\frac{\partial \tilde{\xi}_\alpha}{\partial \varrho}\right)_0 \varrho' + \left(\frac{\partial \tilde{\xi}_\alpha}{\partial \theta}\right)_0 \theta'\right]\mu'_\alpha + O(\varepsilon^3)$$
$$(2.5.40)$$

as $\varepsilon \to 0$. Thus

$$(2.5.41) \qquad \left(\frac{\partial \tilde{\xi}_{,\alpha}}{\partial \varrho}\right)_0 = \left(\frac{\partial \tilde{\xi}_{,\alpha}}{\partial \theta}\right)_0 = 0,$$

and $\|\tau_{\alpha\beta}\|$ must be positive semi-definite. Finally, the Taylor expansion of (2.5.33) about $s_0$ reduces to (2.5.35) when account is taken of (2.5.1), (2.5.37) and (2.5.41).

# REFERENCES

[1940] Eckart, C., The thermodynamics of irreversible processes, II. Fluid mixtures, Phys. Rev. 58, 2 269-275.

[1957] Truesdell, C., Sulle basi della termomeccanica, Rend. Accad. Lincei 22, 33-88, 158-166. English transl. in Rational Mechanics of Materials, Intl. Sci. Rev. Ser. 292-305, Gordon and Breach, New York, 1965.

[1959] Meixner, J. and H. G. Reik, Thermodynamik der irreversiblen Prozesse, The Encyclopedia of Physics. Vol. 111/2, edited by S. Flügge, Springer, Berlin-Gottingen-Heidelberg.

Nachbar, W., F. Williams, and S. S. Penner, The conservation equations for independent coexistent continua and for multicomponent reacting gas mixtures, Quart. Appl. Math. 17, 43-54.

[1962] de Groot, S. R. and P. Mazur, Non-Equilibrium Thermodynamics, Interscience, New York.

[1963] Coleman, B.D. and W. Noll, The thermodynamics of elastic materials with heat conduction and viscosity, Arch. Rational Mech. Anal. 13, 167 - 170.

[1964] Kelly, P. D. A reacting continuum, Int. J. Engng. Sci. 2, 129-153.

Bowen, R. M., Towards a thermodynamics and mechanics of mixtures, Arch. Rational Mech. Anal. 24, 370-403.

Crochet, M. J. and P. M. Naghdi, Small motions superposed on large static deformations in porous media, Acta Mech. 4, 315-335.

Green, A. E. and P.M. Naghdi, A theory of mixtures Arch. Rational Mech. Anal. 24, 243-263.

Ingram, J. D. and A. C. Eringen, A continuum theory of chemically reacting contina, 11. Constitutive equations of reacting fluid mixtures, Int. J. Engng. Sci. 5, 289-322.

Mills, N., A theory of multi-component mixtures, Quart. J. Mech. Appl. Math. 20, 499-508.

[1968] Dunwoody, N. T. and I. Miller, A thermodynamic theory of two chemically reacting ideal gases with different temperatures, Arch. Rational Mech. Anal. 29, 344-369.

Müller, I., A thermodynamic theory of mixtures of fluids, Arch. Rational Mech. Anal. 28, 1-39.

Truesdell, C., Sulle basi della termodinamica delle miscele, Rend. Accad. Naz. Lincei, (8)44, 381-383.

[1969] Bowen, R. M. The thermochemistry of a reacting mixture of elastic materials with diffusion, Arch. Rational Mech. Anal. 34, 97-127.

Bowen, R. M. and J. C. Weise, Diffusion in mixtures of elastic materials, Int. J. Engng, Sci. 7, 689-722.

Doria, M. L., Some general results for non-reacting binary mixtures of fluids, Arch. Rational Mech. Anal. 32, 343-368.

Green, A. E. and P. M. Naghdi, On basic equations for mixtures, Quart. J. Mech. Appl. Math. 22, 427-438.

# References

Truesdell, C. Rational Thermodynamics : <u>A Course of on Selected Topics</u>, McGraw- Hill, New York.

[1971] Gurtin, M. E., On the thermodynamics of chemically fluid mixtures, Arch. Rational Mech. Anal. 43, 198-212.

Gurtin, M. E. and A. S. Vargas, On the classical theory of fluid mixtures, Arch. Rational Mech. Anal. 43, 179-197.

## CONTENTS

|  | Page |
|---|---|
| Preface.................................................. | 3 |
| Introduction............................................ | 5 |
| Chapter I : Thermodynamics of Elastic Materials | |
|    1.1  Basic Laws......................... | 6 |
|    1.2  Elastic Materials................... | 10 |
| Chapter II: Mixtures | |
|    2.1  Mechanics of Mixtures............... | 16 |
|    2.2  Thermodynamics of Mixtures.......... | 21 |
|    2.3  Constitutive Relations.............. | 24 |
|    2.4  Consequences of the Second Law...... | 27 |
|    2.5  Results near Equilibrium............ | 33 |
| References.............................................. | 43 |
| Contents................................................ | 47 |

MIX
Papier aus verantwortungsvollen Quellen
Paper from responsible sources
FSC® C105338

If you have any concerns about our products,
you can contact us on
ProductSafety@springernature.com

In case Publisher is established outside the EU,
the EU authorized representative is:
**Springer Nature Customer Service Center GmbH
Europaplatz 3, 69115 Heidelberg, Germany**

Printed by Libri Plureos GmbH
in Hamburg, Germany